Ready For 4th Grade

Math

A Summer Learning Activity Workbook For Students Entering Grade 4 [Ages 9-10]

Written and edited by Matthew Bowling

Printed in the United States of America

This paper meets the requirements of ANSI/NISO Z39.48-1992 (Permanence of Paper).

Paperback
ISBN 13: 978-1-5167-3138-1
ISBN 10: 1-5167-3138-7

TABLE OF CONTENTS

Introduction

You did it! You finished 3rd grade! That is so exciting and awesome!

You are about to head into 4th grade now, and that may be a little scary. This workbook is here to help you feel ready and confident as you tackle this new grade level. You've got this!

This workbook is organized into chapters. Each chapter will review important 3rd grade skills and introduce new ideas you'll see in 4th grade. Each chapter starts with a short lesson to refresh your memory or even introduce new things before you take on the activities for each day. You can always go back to that page for help when doing an activity page.

The workbook is made to be completed over 10 weeks, with each day taking no more than 15 minutes. That is just enough time to practice without feeling overwhelmed. Learning is an adventure! As you work through the chapters, the activities will slowly become trickier, but that just means your brain is growing stronger! This will help prepare you for bigger thinking and new tasks you will see in 4th grade.

You might work with a grown-up sometimes, or you might complete much of it on your own. Either way, you're building skills, confidence, and a love for learning. Get ready, because 4th grade is going to be a great year!

Chapter 1: Place Value & Number Sense

Numbers can be ordered from least to greatest or greatest to least. Least to greatest means smallest to largest. Greatest to least means largest to smallest. A number line may be useful when ordering numbers.

Rounding is used to change a number to a nearby number that is easier to work with. To round, look at the digit after the place you want to round to. For example, if you are asked to round to the nearest tens place, look at the ones place. If the digit is 0, 1, 2, 3, or 4, the number in the place you are rounding to stays the same. This is referred to as rounding down. For example, 342 rounded to the nearest tens place is 340. If the digit is 5, 6, 7, 8, or 9, the number in the place you are rounding to goes up to the next value. This is referred to as rounding up. For example, 768 rounded to the nearest tens place is 770.

Numbers can be broken apart and composed based on their **place value**. For example, the number 129,548 can be expressed in **expanded form** as 100,000 + 20,000 + 9,000 + 500 + 40 + 8. You should also be comfortable going from expanded form to standard form. For example, what number is represented by 30,000 + 400 + 80 + 2? Notice 3 groups of ten-thousands, 0 groups of thousands, 4 groups of hundreds, 8 groups of tens, and 2 groups of ones. This becomes 30,482.

Sometimes you will be asked to compose and decompose numbers that are presented in **picture** or **model** form. This usually includes base-10 blocks. For example, consider the model below. How many groups of thousands, hundreds, tens, and ones are there? There is 1 group of 1,000, 2 groups of 100, 4 groups of 10, and 7 ones. This forms the number 1,247.

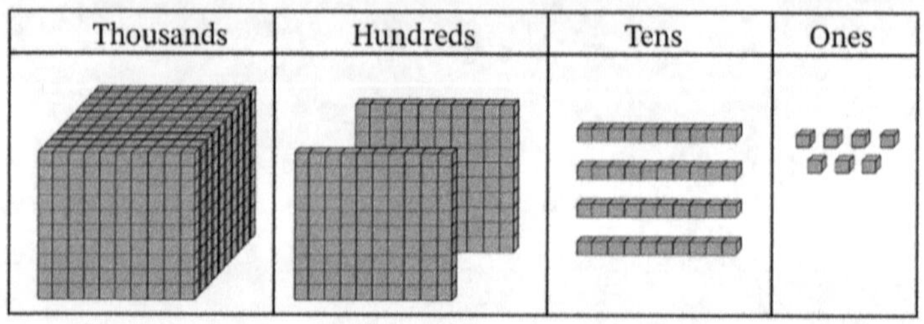

Complete one worksheet each day in this chapter. Each day should take no longer than 15 minutes to complete.

DAY 1

PLACE VALUE: BIG NUMBERS 1-1,000,000

The place value of a digit is determined by where it is in a number.

Millions	Hundred Thousands	Ten Thousands	Thousands	Hundreds	Tens	Ones
1	2	3	4	5	6	7

1, 2 3 4, 5 6 7

<u>One Million, Two Hundred Thirty-Four Thousand, Five Hundred Sixty-Seven</u>

Write these numbers correctly in the blanks below.

1. 6,138,462	=	6 Millions	1 Hundred Thousands	3 Ten Thousands	8 Thousands	4 Hundreds	6 Tens	2 Ones
2. 3,194,675	=	Millions	Hundred Thousands	Ten Thousands	Thousands	Hundreds	Tens	Ones
3. 8,417,205	=	Millions	Hundred Thousands	Ten Thousands	Thousands	Hundreds	Tens	Ones
4. 2,765,447	=	Millions	Hundred Thousands	Ten Thousands	Thousands	Hundreds	Tens	Ones
5. 5,925,057	=	Millions	Hundred Thousands	Ten Thousands	Thousands	Hundreds	Tens	Ones

Day 2

Identifying Place Value: Word Problems

Solve the word problems below.

1. In the number 25,483:

   a. This digit is in the ones place: ________________________.

   b. This digit is in the hundreds place: ________________________.

   c. The five is in the ________________________ place.

   d. The eight is in the ________________________ place.

2. In the number 62,134:

   a. This digit is in the tens place: ________________________.

   b. This digit is in the thousands place: ________________________.

   c. The six is in the ________________________ place.

   d. The four is in the ________________________ place.

3. In the number 84,327:

   a. This digit is in the ones place: ________________________.

   b. This digit is in the thousands place: ________________________.

   c. The two is in the ________________________ place.

   d. The eight is in the ________________________ place.

4. In the number 14,960:

   a. This digit is in the hundreds place: ________________________.

   b. This digit is in the ten-thousands place: ________________________.

   c. The zero is in the ________________________ place.

   d. The four is in the ________________________ place.

5. In the number 40,589:

   a. This digit is in the ones place: ________________________.

   b. This digit is in the hundreds place: ________________________.

   c. The zero is in the ________________________ place.

   d. The four is in the ________________________ place.

Day 3

Rounding Numbers

Round the following to the nearest thousand.

1. 2,563
2. 4,219
3. 9,198
4. 5,756
5. 1,423
6. 8,154

Round the following to the nearest ten-thousand.

7. 39,092
8. 77,150
9. 19,917
10. 33,809
11. 93,254
12. 35,451

Round the following to the nearest hundred-thousand.

13. 274,333
14. 317,110
15. 596,559
16. 882,658
17. 221,324
18. 610,567

Day 4

Ordering up to 100,000

Write these numbers in order from least to greatest.

1.	100,289	100,891	100,280	100,381
2.	512,112	512,123	51,110	512,101
3.	855,622	865,622	856,628	856,629
4.	634,209	634,101	635,700	633,601
5.	231,115	20,115	231,150	21,115
6.	546,029	546,209	546,030	546,020
7.	375,787	355,891	355,111	375,375
8.	23,151	239,151	237,151	240,151
9.	155,289	146,891	145,280	145,381
10.	269,722	20,722	269,772	266,722

Chapter 2: Addition

Addition is when two or more sets of things are combined or joined together. This is also referred to as **composing**. The **sum** is the answer achieved after adding the sets. Similar to counting, unique symbols are used to indicate addition. The **plus sign** (+) represents addition, and the **equal sign** (=) is used to indicate the sum or answer. Numbers are used to represent the different sets of things that are being combined.

Addition can be done using numbers, visuals, and physical objects.

To add multiple digit numbers together, start in the ones place and use basic addition rules. When the number equals ten or more, the first digit of the number carries over to the next place value. This is called **regrouping**. Look at the example below. When the numbers in the ones column add up to a number larger than 10, the first digit of that number carries over to the tens column.

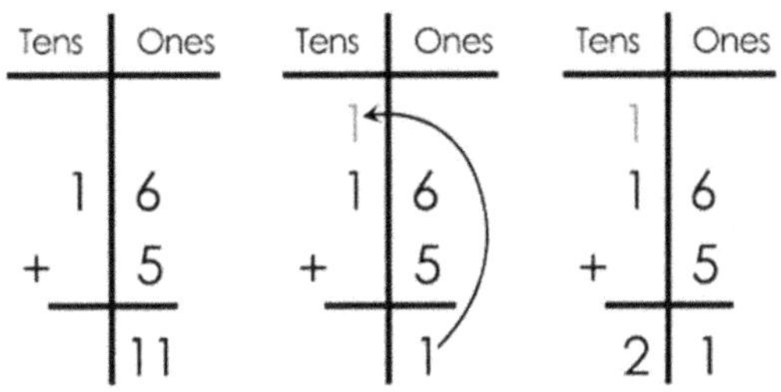

This same process is used for larger numbers as well.

Complete one worksheet each day in this chapter. Each day should take no longer than 15 minutes to complete.

Day 5

2-Digit 3-Row Addition: Regrouping

Solve the two-digit 3-row addition problems below.

1.	2.	3.	4.
3 6 2 2 + 1 5 ____	8 4 7 0 + 9 ____	6 4 2 9 + 4 2 ____	5 7 5 2 + 7 3 ____
5.	**6.**	**7.**	**8.**
4 4 1 5 + 3 9 ____	6 6 9 9 + 8 2 ____	5 0 3 6 + 5 5 ____	1 5 3 5 + 8 0 ____
9.	**10.**	**11.**	**12.**
6 7 7 2 + 1 8 ____	1 1 2 8 + 3 7 ____	6 2 9 7 + 1 9 ____	5 5 2 0 + 1 2 ____

Day 6

Addition Word Problems

1. Elle picked apples for four days. On day one she picked 10 apples. On day two she picked 26 apples. On day three she picked 14 apples. On day four she picked 14 apples. How many apples does Elle have in total?

____

2. Cindy has been doing her chores every day after school. On Monday she swept the floor for 12 minutes. On Tuesday she spent 9 minutes making her bed. On Wednesday she washed the dishes for 20 minutes. On Thursday she spent 15 minutes vacuuming the floor. How much time did she spend on her chores?

____

3. Steven is good at basketball. In game one he scored 14 points. In game two he scored 22 points. In game three he scored 27. In games four and five he scored 17 and 12 points. How many points has Steven scored so far?

____

4. Amy loves to read. She reads every day after school. One day she read 45 minutes. The next day she read 30 minutes. The day after that she spent 57 minutes reading. Yesterday she read for 26 minutes and today she spent 15 minutes reading her book. How much time has she spent reading this week?

____

Day 7

3-Digit Addition: Regrouping 1

Solve the 3-digit addition problems below.

1.	2.	3.	4.
1 4 9 8 + 3 2 1 8 1 9	5 8 0 + 7 2 9	1 3 4 + 2 2	3 0 9 + 4 4 7
5.	6.	7.	8.
2 5 3 + 2 0 3	1 7 1 + 8 2	3 4 4 + 4 9 3	7 1 4 + 5 0 7

Day 8

3-Digit Addition: Regrouping 2

Solve the 3-digit addition problems below.

1.	2.	3.	4.
carries: 1 1 151 + 459 610	239 + 205	677 + 380	406 + 312
5. 951 + 123	6. 705 + 357	7. 824 + 199	8. 901 + 699
9. 333 + 178	10. 101 + 752	11. 509 + 456	12. 712 + 55

Chapter 2: Addition

Day 9

4-Digit Addition: Regrouping

Solve the 4-digit addition problems below.

1.	2.	3.	4.
1 1 1 1 6, 8 9 5 + 5, 4 0 6 ——— 1 2, 3 0 1	1, 2 5 9 + 9, 5 0 2 ———	9, 5 4 3 + 7, 8 4 5 ———	4, 0 5 1 + 1, 1 5 0 ———
5.	**6.**	**7.**	**8.**
8, 5 0 7 + 9, 8 4 7 ———	3, 2 5 0 + 4 0 6 ———	2, 4 7 0 + 3, 3 5 7 ———	1, 5 8 9 + 6, 8 7 5 ———

Day 10

5-Digit Addition: Regrouping

Solve the 5-digit addition problems below.

1.	2.	3.	4.
15,801 + 35,847 51,648	75,211 + 25,786	89,057 + 11,059	33,520 + 6,579
5.	6.	7.	8.
40,517 + 12,523	74,259 + 69,805	94,521 + 7,089	30,671 + 31,337

Chapter 3: Subtraction

Subtraction is when a set of things has a portion that is separated, removed, or taken away. This is also referred to as **decomposing**. The **difference** is the answer achieved after subtracting from the set. Similar to counting and addition, unique symbols are used to indicate subtraction. The **minus sign** (-) represents subtraction, and the **equal sign** (=) is used to indicate the difference or answer. Numbers are used to represent the different amounts.

Subtraction can be done using numbers, visuals, and physical objects.

To subtract multiple digit numbers together, start in the ones place and use basic subtraction rules. If the top number in the column is smaller than the bottom, borrow a group of ten from the column to the left. This is called **borrowing**. Look at the example below.

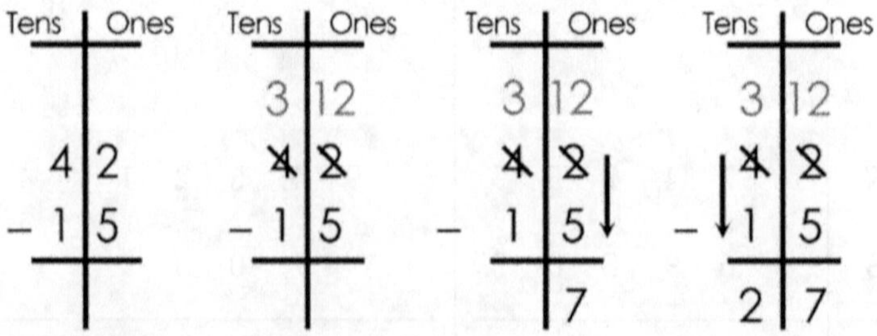

This same process is used for larger numbers as well.

Complete one worksheet each day in this chapter. Each day should take no longer than 15 minutes to complete.

Day 11

2-Digit Subtraction: Borrowing

Solve the two-digit subtraction problems below.

2 15 ~~3~~ ~~5~~ - 6	2. 5 4 - 2 8	3. 8 3 - 4	4. 4 2 - 1 5
5. 5 6 - 9	6. 4 1 - 5	7. 8 2 - 5 9	8. 1 6 - 7

Day 12

Subtraction Word Problems

1. Selena is a good soccer player. This season she has taken 38 shots at the goal and made 6 of them. How many shots has she missed this season?

____

2. Billy can't stop eating cookies because they are so yummy. He had 62 cookies in his jar. He ate 18 of them. How many cookies are left in the jar?

____

3. Aida loves baseball. She has gone to 103 games in her life. 39 were away games. How many were home games?

____

4. Jonathan has collected 47 snowflakes today, but 23 of them have already melted. How many snowflakes does Jonathan have left?

____

DAY 13

3-DIGIT SUBTRACTION: BORROWING

Solve the 3-digit subtraction problems below.

1. 8 9 16 ~~9~~ ~~0~~ ~~6~~ - 6 8 7 2 1 9	2. 4 1 2 - 3 8 9	3. 5 1 8 - 7 9	4. 9 6 2 - 4 7 3
5. 4 1 2 - 2 7 3	6. 6 5 2 - 3 8 6	7. 1 7 9 - 8 3	8. 7 1 2 - 5 5 4

Day 14

4-Digit Subtraction: Borrowing

Solve the 4-digit subtraction problems below.

1.	2.	3.	4.
1 14 7 13 ~~2~~, ~~4~~ ~~6~~ ~~3~~ - 1, 9 3 9 ——— 5 4 4	2, 6 5 5 - 1, 5 6 4 ———	4, 2 3 3 - 4 5 2 ———	5, 4 2 8 - 2, 6 4 9 ———
5.	**6.**	**7.**	**8.**
1, 9 9 5 - 2 3 9 ———	7, 3 2 1 - 8 3 2 ———	9, 2 1 1 - 1, 7 0 0 ———	3, 9 4 6 - 1, 6 8 2 ———

Day 15

5-Digit Subtraction: Borrowing

Solve the 5-digit subtraction problems below.

1.	2.
5 12 6 12 14 ~~6~~ ~~2~~, ~~7~~ ~~3~~ 4 - 4 9, 5 8 6 1 3, 1 4 8	4 5, 6 8 9 - 2 9, 8 7 5
3.	**4.**
9 0, 2 5 4 - 6 7, 7 0 2	1 2, 0 9 7 - 7, 8 5 0
5.	**6.**
8 0, 7 7 5 - 2 4, 0 7 8	3 7, 8 2 0 - 9, 5 0 7
7.	**8.**
5 4, 5 1 9 - 2 0, 1 7 9	7 0, 4 2 1 - 3 4, 2 1 9

Chapter 3: Subtraction

Chapter 4: Decimals

The important thing to remember when adding and subtracting decimal numbers is to **line up** the **decimal points**. Lining up the decimal points ensures that similar place values are being combined. For example, ones combine with ones, tens combine with tens.

Consider the following: A pizza costs $27.50 and a drink costs $4.65. The total cost of the meal can be calculated by finding the sum. Stack the numbers vertically and line up the decimal points. With addition, it does not matter which number is on top or bottom. Regroup if necessary and bring the decimal straight down into the answer line. This method is known as the **standard algorithm**.

```
  11
  27. 50
+  4. 65
$32. 15
```

The standard algorithm is also useful for subtraction scenarios. For example, if a piece of ribbon measures 132.1 centimeters, and 27.76 centimeters are cut off and removed, how many centimeters of ribbon are left?

Stack the values vertically and line up the decimals. Make sure the larger number, 132.1, is on the top. Sometimes it is helpful to fill in any empty spaces with a zero. Regroup if necessary, and again bring the decimal straight down into the answer line.

```
      11   10
     2 1   0 10
   1 3 2 . 1 0
-    2 7 . 7 6
   1 0 4 . 3 4
```

There are 104.34 centimeters of ribbon left.

Money is expressed in **dollars** and **cents**. Cents are represented as **tenths** and **hundredths**. For example, $4.92 represents 4 whole dollars, 9 tenths of a dollar, and 2 hundredths of a dollar. These same rules are used to add or subtract money.

Decimal values fall **between two whole numbers** on a **number line**. For example, 5.7 falls between 5 and 6 on a number line. When rounding a decimal to a whole number, look for the whole number it is closest to on the number line. 5.7 is closer to 6 than 5, so 5.7 rounds to 6. Decimals will not always need to be rounded to the nearest whole number. Decimals can also be rounded to the nearest tenth or hundredth.

Complete one worksheet each day in this chapter. Each day should take no longer than 15 minutes to complete.

DAY 16

ADDING DECIMALS

Line up the numbers and solve the problems below. Be sure to show your work.

1. 1.21 + 2.33 + 0.05 1 . 2 1 2 . 3 3 + 0 . 0 5 3 . 5 9	2. 4 + 1.63 + 2.2 . . + . .	3. 0.24 + 4.87 + 11 . . + . .
4. 7.12 + 2.04 + 5 . . + . .	5. 9.11 + 0.43 + 0.89 . . + . .	6. 64 + 8.79 + 1.57 . . + . .

Day 17

Adding Money

Find the totals below.

1.	2.	3.
$\begin{array}{r} {}^{1} \\ \$\ 1.40 \\ +\ \$\ 3.63 \\ \hline \$\ 4.03 \end{array}$	$\begin{array}{r} \$\ 7.18 \\ +\ \$\ 4.45 \\ \hline \$\ . \end{array}$	$\begin{array}{r} \$\ 2.50 \\ +\ \$\ 6.72 \\ \hline \$\ . \end{array}$
4.	**5.**	**6.**
$\begin{array}{r} \$\ 1.15 \\ +\ \$\ 5.25 \\ \hline \$\ . \end{array}$	$\begin{array}{r} \$\ 9.40 \\ +\ \$\ 3.35 \\ \hline \$\ . \end{array}$	$\begin{array}{r} \$\ 10.45 \\ +\ \$\ 11.82 \\ \hline \$\ . \end{array}$
7.	**8.**	**9.**
$\begin{array}{r} \$\ 20.15 \\ +\ \$\ 2.38 \\ \hline \$\ . \end{array}$	$\begin{array}{r} \$\ 14.05 \\ +\ \$\ 5.59 \\ \hline \$\ . \end{array}$	$\begin{array}{r} \$\ 12.35 \\ +\ \$\ 2.78 \\ \hline \$\ . \end{array}$

Day 18

Adding Decimals Word Problems

1. Jimmie's water gun can hold 30.08 ounces of water. Steven's water gun can hold 22.56 ounces. How many ounces of water do they have together?

___

2. Sara walks every afternoon. On Monday she walked 2.3 miles. On Tuesday she walked 1.8 miles. On Wednesday she walked 3.1 miles. How many miles did she walk in total?

___

3. Mickey loves to race his bike. In race one he had a finishing time of 13.38 minutes. In race two he had a finishing time of 12.58 minutes. In race three he had a finishing time of 12.32 minutes. What was his total time for all the races?

___

4. Mary can carry 6.75 pounds of dirt in her wagon. Mark can carry 8.25 pounds in his wagon. How much dirt can they carry in total?

___

Day 19

Rounding Decimals

Round each decimal to the nearest whole number.

1. 92.8 =______________	11. 34.2 =______________
2. 12.5 =______________	12. 31.8 =______________
3. 33.1 =______________	13. 76.7 =______________
4. 24.9 =______________	14. 72.2 =______________
5. 57.7 =______________	15. 12.3 =______________
6. 65.1 =______________	16. 43.7 =______________
7. 82.5 =______________	17. 45.8 =______________
8. 41.6 =______________	18. 55.9 =______________
9. 68.6 =______________	19. 21.4 =______________
10. 20.4 =______________	20. 19.3 =______________

Day 20

Subtracting Decimals

Line up the numbers and solve the problems below. Be sure to show your work.

1. 6.05 – 3.89 5 . 9 15 ~~6~~ . ~~0~~ ~~5~~ - 3 . 8 9 2 . 1 6	2. 48.27 – 13.65 . - . .	3. 8.55 – 3.9 . - . .
4. 331 – 4.72 . - . .	5. 902.22 – 47.79 . - . .	6. 75.5 – 6.51 . - . .

Chapter 4: Decimals

Day 21

Subtracting Money

Find the differences below.

1.	2.	3.
$4.58 - $2.11 $2.47	$8.49 - $5.35 $.	$6.50 - $3.32 $.
4.	5.	6.
$9.75 - $7.89 $.	$9.90 - $8.75 $.	$2.93 - $1.08 $.
7.	8.	9.
$13.04 - $8.25 $.	$31.35 - $9.07 $.	$15.89 - $2.35 $.

Day 22

Subtracting Decimals Word Problems

1. Pete wants a new teddy bear. His mom is going to help him buy one. A new bear costs $19.50. Pete has $7.32. How much money does he need to get from mom?

____

2. Tammy wants a new jump rope. She has $3.78. The jump rope cost $6.99. How much more money does she need?

____

3. Randy and John want a new toy truck. The toy truck costs $413.75. Randy has $4.15 and John has $6.42. How much more money do they need?

____

4. Eric sold his old pogo stick for $5.89. A new one costs $35.99. How much more money does Eric need?

____

Chapter 5: Fractions

Fractions are numbers that make up part of a whole.

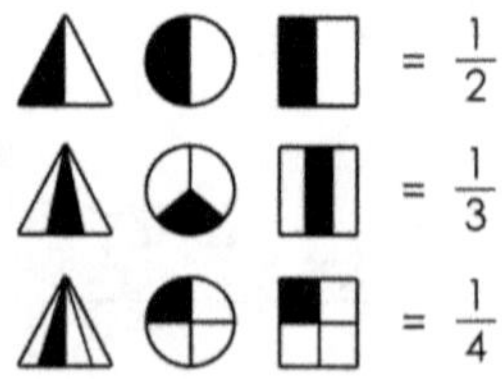

The shaded parts represent the top of the fractions, which are called **numerators**. The bottom numbers, **denominators**, represent the total number of equal pieces each shape is broken up into.

Mixed numbers are values that are written using a whole number and a fraction. For example, $2\frac{1}{3}$ is a mixed number that consists of 7 groups of $\frac{1}{3}$. **Improper fractions** are fractions that have a numerator that is greater than the denominator. $\frac{4}{3}$ is an improper fraction that consists of 4 groups of $\frac{1}{3}$.

A picture or number line can be helpful to compare fractions, but it is not always necessary. Consider two fractions that have the same denominator. The two fractions $\frac{2}{4}$ and $\frac{3}{4}$ both have 4 in the denominator. This means both fractions have a whole of 4. However, 2 parts out of 4 is less than 3 parts out of 4. Think of it this way: 2 pieces are less than 3 pieces, so $\frac{2}{4}$ is less than $\frac{3}{4}$: $\frac{2}{4} < \frac{3}{4}$.

Fractions with the same denominator can be combined by adding the parts (the numerators, or top numbers). This can be represented using a model. For example, the model below shows 3 parts out of 6, and 1 part out of 6. Both fractions consist of sixths, so the parts can be combined. 3 parts combined with 1 part equals 4 parts. 4 parts out of 6 parts represents the fraction $\frac{4}{6}$. In other words, the whole model shows $\frac{3}{6} + \frac{1}{6} = \frac{4}{6}$.

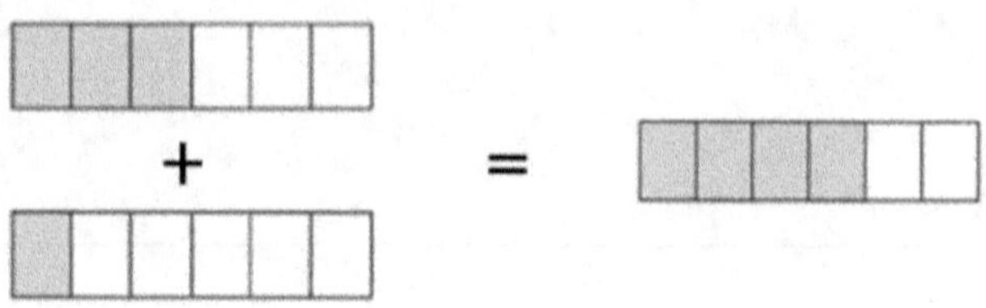

Fractions with the same denominator can be subtracted in the same way. The model below shows 3 parts out of 6, and 1 part out of 6. Both fractions consist of sixths, so the parts can be subtracted. 1

part taken away from 3 parts equals 2 parts. 2 parts out of 6 parts represents the fraction $\frac{2}{6}$. In other words, the whole model shows $\frac{3}{6} - \frac{1}{6} = \frac{2}{6}$.

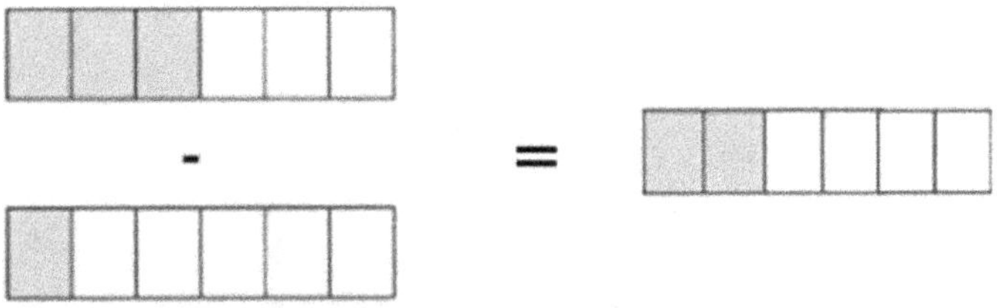

Simplifying fractions means reducing a fraction to the lowest possible equivalent term. To do this, find the greatest common factor (GCF) of the numerator and the denominator. Then divide both the numerator and denominator by that factor.

Complete one worksheet each day in this chapter. Each day should take no longer than 15 minutes to complete.

DAY 23
WRITING FRACTIONS

Write what fraction of each set is shaded in.

1.

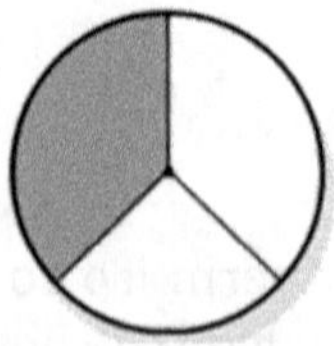

= $\frac{1}{3}$

2.

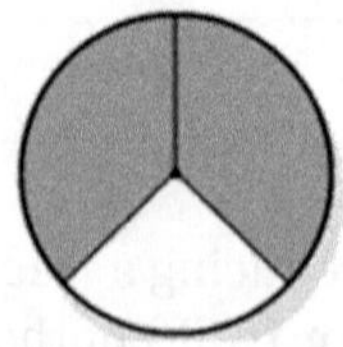

= ____

3.

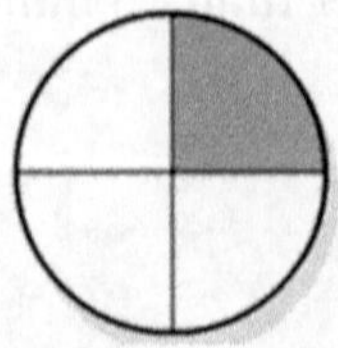

= ____

4.

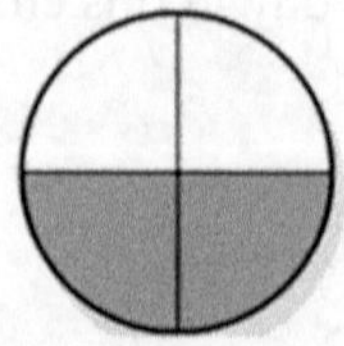

= ____

5.

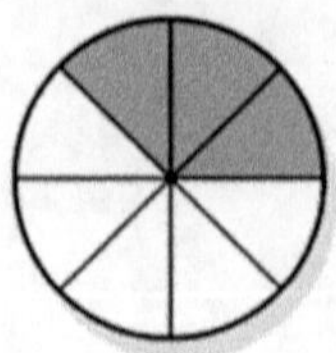

= ____

6.

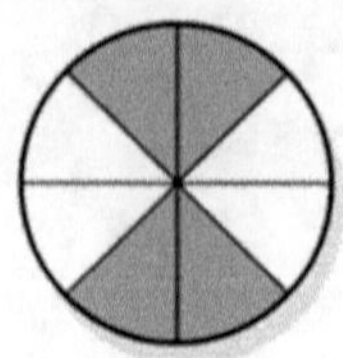

= ____

7.

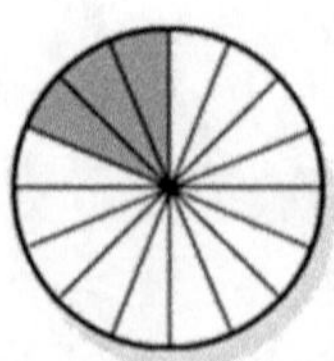

= ____

8.

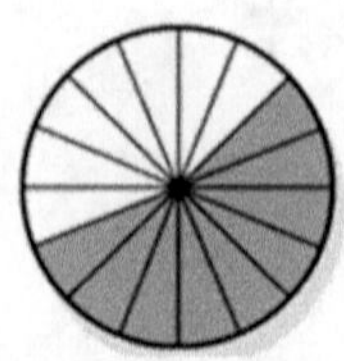

= ____

DAY 24

COMPARING FRACTIONS

Compare the fractions using the symbols >, <, or =.

1. $\frac{4}{9}$ [>] $\frac{3}{9}$
2. $\frac{5}{8}$ [ ] $\frac{7}{8}$
3. $\frac{3}{4}$ [ ] $\frac{1}{4}$
4. $\frac{1}{2}$ [ ] $\frac{2}{2}$
5. $\frac{7}{9}$ [ ] $\frac{5}{9}$
6. $\frac{6}{40}$ [ ] $\frac{16}{40}$
7. $\frac{10}{13}$ [ ] $\frac{12}{13}$
8. $\frac{6}{7}$ [ ] $\frac{5}{7}$
9. $\frac{9}{16}$ [ ] $\frac{2}{16}$
10. $\frac{10}{11}$ [ ] $\frac{2}{11}$
11. $\frac{2}{3}$ [ ] $\frac{2}{3}$
12. $\frac{25}{52}$ [ ] $\frac{4}{52}$
13. $\frac{60}{78}$ [ ] $\frac{9}{78}$
14. $\frac{1}{10}$ [ ] $\frac{2}{10}$
15. $\frac{2}{3}$ [ ] $\frac{4}{3}$
16. $\frac{1}{12}$ [ ] $\frac{2}{12}$
17. $\frac{1}{9}$ [ ] $\frac{2}{9}$
18. $\frac{23}{29}$ [ ] $\frac{4}{29}$
19. $\frac{4}{7}$ [ ] $\frac{6}{7}$
20. $\frac{111}{12}$ [ ] $\frac{12}{12}$

Day 25

Ordering Mixed Numbers

Write the mixed numbers in order from least to greatest.

1. $1\frac{6}{10}, 2\frac{3}{10}, 1\frac{1}{10}, 2\frac{1}{10}$ $1\frac{1}{10}, 1\frac{6}{10}, 2\frac{1}{10}, 2\frac{3}{10}$

2. $4\frac{4}{8}, 9\frac{6}{8}, 6\frac{2}{8}, 9\frac{3}{8},$

3. $6\frac{1}{4}, 7\frac{1}{4}, 6\frac{2}{4}, 7\frac{3}{4}$

4. $8\frac{1}{9}, 9\frac{1}{9}, 8\frac{2}{9}, 8\frac{3}{9}$

5. $3\frac{8}{12}, 3\frac{1}{12}, 1\frac{1}{12}, 2\frac{3}{12}$

6. $5\frac{3}{54}, 5\frac{2}{54}, 5\frac{1}{54}, 5\frac{7}{54}$

Day 26

Adding Fractions with Common Denominators

Add the fractions below.

1. $\frac{1}{5}+\frac{3}{5}=$ $\frac{4}{5}$ ______

2. $\frac{5}{10}+\frac{3}{10}=$ ______

3. $\frac{3}{6}+\frac{2}{6}=$ ______

4. $\frac{7}{12}+\frac{3}{12}=$ ______

5. $\frac{4}{9}+\frac{4}{9}=$ ______

6. $\frac{6}{14}+\frac{2}{14}=$ ______

7. $\frac{3}{7}+\frac{2}{7}=$ ______

8. $\frac{1}{4}+\frac{2}{4}=$ ______

9. $\frac{2}{6}+\frac{3}{6}=$ ______

10. $\frac{4}{8}+\frac{3}{8}=$ ______

11. $\frac{6}{11}+\frac{3}{11}=$ ______

12. $\frac{7}{15}+\frac{5}{15}=$ ______

13. $\frac{3}{14}+\frac{9}{14}=$ ______

14. $\frac{1}{3}+\frac{1}{3}=$ ______

15. $\frac{2}{8}+\frac{4}{8}=$ ______

16. $\frac{3}{5}+\frac{1}{5}=$ ______

17. $\frac{9}{13}+\frac{2}{13}=$ ______

18. $\frac{1}{10}+\frac{6}{10}=$ ______

Day 27

Subtracting Fractions with Common Denominators

Subtract the fractions below.

1. $\frac{7}{15} - \frac{3}{15} =$ $\frac{4}{15}$ ______
2. $\frac{7}{9} - \frac{2}{9} =$ ______
3. $\frac{12}{14} - \frac{7}{14} =$ ______
4. $\frac{8}{8} - \frac{6}{8} =$ ______
5. $\frac{28}{39} - \frac{19}{39} =$ ______
6. $\frac{24}{39} - \frac{11}{39} =$ ______
7. $\frac{9}{10} - \frac{2}{10} =$ ______
8. $\frac{13}{15} - \frac{9}{15} =$ ______
9. $\frac{13}{16} - \frac{3}{16} =$ ______
10. $\frac{6}{7} - \frac{4}{7} =$ ______
11. $\frac{10}{26} - \frac{8}{26} =$ ______
12. $\frac{60}{62} - \frac{41}{62} =$ ______
13. $\frac{19}{24} - \frac{10}{24} =$ ______
14. $\frac{46}{54} - \frac{17}{54} =$ ______
15. $\frac{98}{98} - \frac{15}{98} =$ ______

Day 28

Simplifying Fractions

Finish writing the problems out to reduce the fractions to their simplest form.

1. $\frac{5}{15} \div \frac{3}{3} = \frac{1}{3}$

2. $\frac{4}{12} \div$ ______ = ______

3. $\frac{6}{15} \div$ ______ = ______

4. $\frac{8}{10} \div$ ______ = ______

5. $\frac{10}{25} \div$ ______ = ______

6. $\frac{20}{25} \div$ ______ = ______

Chapter 6: Multiplication

Multiplication is the process of counting equally sized groups. The multiplication sign (×) is read as times. For example, 5 × 7 is read as five times seven. This means 5 groups of 7 or 7 + 7 + 7 + 7 + 7. The answer to a multiplication problem is called the **product**. Thus, the product of 5 groups of 7 is 35. The process of using addition to solve multiplication problems is known as **repeated addition**.

Factors are the numbers that are being multiplied together. For example, the factors in the multiplication problem 4 × 3 = 12 are four and three. Below shows a standard multiplication table with factors and products.

x	1	2	3	4	5	6	7	8	9	10
1	1	2	3	4	5	6	7	8	9	10
2	2	4	6	8	10	12	14	16	18	20
3	3	6	9	12	15	18	21	24	27	30
4	4	8	12	16	20	24	28	32	36	40
5	5	10	15	20	25	30	35	40	45	50
6	6	12	18	24	30	36	42	48	54	60
7	7	14	21	28	35	42	49	56	63	70
8	8	16	24	32	40	48	56	64	72	80
9	9	18	27	36	45	54	63	72	81	90
10	10	20	30	40	50	60	70	80	90	100

An **array** is a collection of **rows** and **columns** that can represent a multiplication scenario.

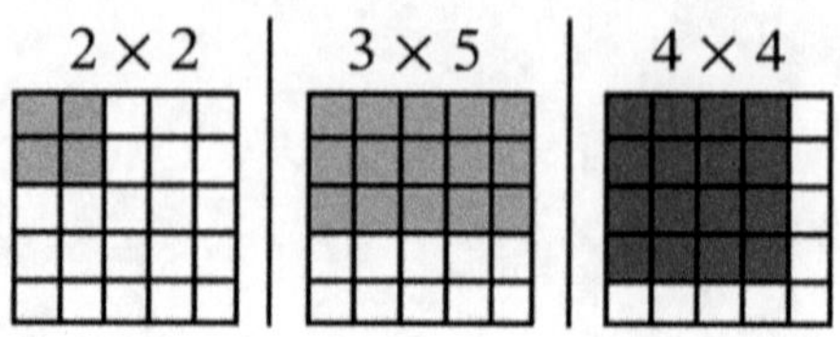

2 rows of 2 columns: 2 × 2 = 4

3 rows of 5 columns: 3 × 5 = 15

4 rows of 4 columns: 4 × 4 = 16

Arrays are great strategies for one and two-digit multiplication. However, the **standard algorithm** is a commonly preferred strategy for numbers that are more than three digits. The standard algorithm, or "vertical multiplication," is less visual but more time-effective. For example, consider the product 7,482 × 8. Write the longer number on the top and the shorter number on the bottom.

Push all numbers to the right so that the place value is lined up (ones lined up with ones, tens lined up with tens, etc.).

$$\begin{array}{r} {}^{3}\ {}^{6}{}^{1}\ \ \\ 7{,}482 \\ \times \quad\quad 8 \\ \hline 59{,}856 \end{array}$$

Distribute, or multiply, the 8 by each digit in the number 7,482. Work from right to left. 8 × 2, then 8 × 8, then 8 × 4, then 8 × 7. If a product has more than one digit, write the ones digit in the answer and carry the tens digit above the next number you are multiplying. For example, 2 × 8 = 16, so write 6 in the answer and 1 above 8 (the next number you will multiply by). Then, add the carried number to the next product. 8 × 8 = 64, then add 1 to get 65, and continue the multiplication process for the rest of the number. The product of 7,482 and 8 is 59,856.

To multiply two multi-digit numbers together, the process is very similar. Start by lining up the ones digit for both numbers and stack the two numbers vertically with the larger number on top. First multiply the digit in the ones place of the bottom number by all digits in the top number from right to left. Then, write a "place holder zero" beneath the initial product in the ones place. Repeat the process by multiplying the digit in the tens place of the bottom number by all the digits in the top number from right to left. Finally, add both products together to get the final product.

$$\begin{array}{r} 1\ \ \\ 526 \\ \times \quad 31 \\ \hline {}^{1}_{1}526 \\ +\ 15780 \\ \hline 16{,}306 \end{array}$$

Complete one worksheet each day in this chapter. Each day should take no longer than 15 minutes to complete.

DAY 29

MULTIPLICATION TABLE PRACTICE

Finish filing out the multiplication tables by multiplying the numbers in the top row by the numbers in the side column.

1.

X	1	2	3	4	5
6					
7					
8					

2.

X	7	8	9
4			
5			
6			
7			
8			

3.

X	9	10	11
7			
8			
9			
10			
11			

4.

X	2	3	4	5	6
20					
25					
27					

Day 30

Mixed Multiplication Facts

Find the product.

1.	2.	3.	4.	5.	6.	7.
2 × 0	8 × 2	1 × 1	9 × 1	2 × 5	3 × 4	9 × 3
8.	**9.**	**10.**	**11.**	**12.**	**13.**	**14.**
4 × 5	3 × 2	7 × 5	2 × 2	3 × 8	6 × 7	6 × 4
15.	**16.**	**17.**	**18.**	**19.**	**20.**	**21.**
2 × 9	5 × 5	1 × 7	2 × 7	1 × 8	5 × 2	8 × 2
22.	**23.**	**24.**	**25.**	**26.**	**27.**	**28.**
6 × 3	6 × 6	1 × 9	5 × 9	3 × 3	6 × 8	9 × 9

Day 31

Multiplying 1-Digit Numbers by 2-Digit Numbers

Find the product.

1.	2.	3.	4.	5.
24 × 2 = 48	31 × 3 =	43 × 2 =	11 × 7 =	11 × 5 =
6.	**7.**	**8.**	**9.**	**10.**
44 × 2 =	15 × 1 =	54 × 2 =	43 × 3 =	34 × 2 =

Day 32

Multiplying by 10s

Find the product.

1.	2.	3.	4.	5.
26 × 10 = 260	71 × 10 =	67 × 10 =	22 × 10 =	38 × 10 =
6.	**7.**	**8.**	**9.**	**10.**
84 × 10 =	19 × 10 =	18 × 10 =	94 × 10 =	29 × 10 =

Day 33

Multiplying 1-Digit Numbers by 2-Digit Numbers: Regrouping

Find the product.

1.	2.	3.	4.	5.
$\begin{array}{r} {}^{1} \\ 46 \\ \times\ 2 \\ \hline 92 \end{array}$	$\begin{array}{r} 58 \\ \times\ 7 \\ \hline \end{array}$	$\begin{array}{r} 43 \\ \times\ 4 \\ \hline \end{array}$	$\begin{array}{r} 87 \\ \times\ 3 \\ \hline \end{array}$	$\begin{array}{r} 36 \\ \times\ 5 \\ \hline \end{array}$
6.	7.	8.	9.	10.
$\begin{array}{r} 44 \\ \times\ 5 \\ \hline \end{array}$	$\begin{array}{r} 32 \\ \times\ 9 \\ \hline \end{array}$	$\begin{array}{r} 55 \\ \times\ 9 \\ \hline \end{array}$	$\begin{array}{r} 99 \\ \times\ 2 \\ \hline \end{array}$	$\begin{array}{r} 59 \\ \times\ 6 \\ \hline \end{array}$

Day 34

Multiplying 1-Digit Numbers by 3-Digit Numbers: Regrouping

Find the product.

1.	2.	3.	4.
$\begin{array}{r} {}^{1} \\ 142 \\ \times 4 \\ \hline 568 \end{array}$	$\begin{array}{r} 221 \\ \times 6 \\ \hline \end{array}$	$\begin{array}{r} 341 \\ \times 3 \\ \hline \end{array}$	$\begin{array}{r} 121 \\ \times 7 \\ \hline \end{array}$
5.	**6.**	**7.**	**8.**
$\begin{array}{r} 698 \\ \times 6 \\ \hline \end{array}$	$\begin{array}{r} 252 \\ \times 4 \\ \hline \end{array}$	$\begin{array}{r} 138 \\ \times 3 \\ \hline \end{array}$	$\begin{array}{r} 239 \\ \times 2 \\ \hline \end{array}$
9.	**10.**	**11.**	**12.**
$\begin{array}{r} 232 \\ \times 7 \\ \hline \end{array}$	$\begin{array}{r} 141 \\ \times 8 \\ \hline \end{array}$	$\begin{array}{r} 212 \\ \times 5 \\ \hline \end{array}$	$\begin{array}{r} 162 \\ \times 3 \\ \hline \end{array}$

Day 35

Multiplying 2-Digit Numbers by 2-Digit Numbers: Regrouping

Find the product.

1.	2.	3.	4.	5.
4 4 9 × 1 5 2 4 5 + 4 9 0 7 3 5	1 3 × 1 2 +	5 4 × 2 2 +	7 2 × 4 5 +	3 8 × 2 5 +
6.	**7.**	**8.**	**9.**	**10.**
8 4 × 5 5 +	6 9 × 3 7 +	2 9 × 2 3 +	7 9 × 5 6 +	9 9 × 5 6 +

Chapter 7: Division

Division is a way to find out how many times one number is counted in another number. Another way of putting this is how many equal groups can be made out of the given amount. For example, 28 bowlers can be divided evenly into 7 groups of 4 people.

This kind of problem can be solved using a variety of strategies. A sketch or visual can be made to evenly partition the bowlers into groups. Multiplication can also be used to determine the answer to the division problem or **quotient**. Similar to addition and subtraction, multiplication and division are **inverse operations**. For example, $72 \div 9$ can be solved using the multiplication facts for 9. $72 \div 9 = ?$ Ask the question, "9 groups of what equals 72?" $9 \times 6 = 54$ ... $9 \times 7 = 63$ ... $9 \times 8 = 72$. The missing number is 8. This means $72 \div 9 = 8$. This strategy is quite helpful because once the multiplication facts are automatic, division facts become automatic as well.

The **division sign** ($\div$ or $\overline{)\ }$) means "divided by." The **dividend** is the number that is divided or broken up into parts. The **divisor** is the number that tells how many parts the dividend will be broken up into.

The problem to the left is read as "six divided by two is equal to three," and the problem to the right is read as "ten divided by two is equal to five."

Sometimes groups of objects or numbers cannot be divided into equal groups, and the number left over in a division problem is called the **remainder**. The remainder must be smaller than the divisor. For example, if 13 is divided into 5 equal groups, there will be 2 equal groups with 3 left over.

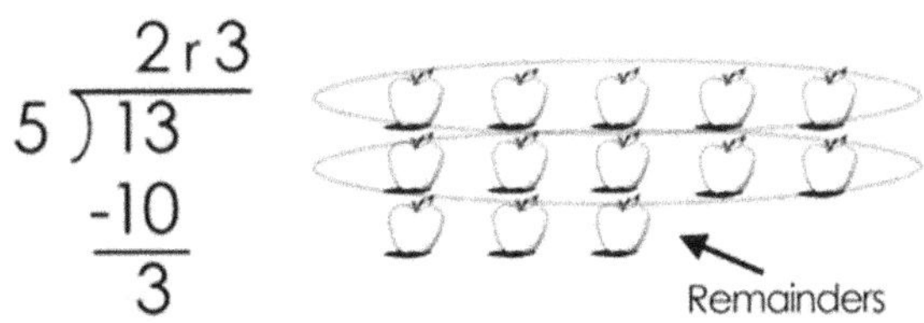

The **standard algorithm** for division, also known as **long division**, is a great option for division problems involving large numbers. To complete long division, the standard process is to estimate, divide, multiply, subtract, compare, bring down, and repeat as necessary. This means start by looking at the first digit in the dividend, and estimating how many times the divisor can evenly fit into the number. For example, 5 cannot fit into 1 at all, so the space above the 1 is left blank. A zero can also be used as a placeholder if needed. Next, 5 can fit into 13 two times evenly. Thus, a 2 is placed above the 3. Since 2 times 5 is 10, place 10 beneath the 13 and subtract. The remaining amount is 3, and there are no other digits that can be brought down to continue the process.

Complete one worksheet each day in this chapter. Each day should take no longer than 15 minutes to complete.

Day 36

Basic Division

Find the quotient.

1. $7\overline{)21}$ with quotient 3

2. $6\overline{)48}$

3. $2\overline{)12}$

4. $7\overline{)28}$

5. $4\overline{)36}$

6. $3\overline{)21}$

7. $7\overline{)21}$

8. $5\overline{)40}$

9. $5\overline{)15}$

10. $2\overline{)6}$

11. $8\overline{)72}$

12. $10\overline{)100}$

Day 37

Division with Remainders

Find the quotient of each problem below.

1.

```
     6 r 1
3 ) 19
   -18
     1
```

2. $7 \overline{)36}$

3. $6 \overline{)25}$

4. $8 \overline{)39}$

5. $5 \overline{)22}$

6. $9 \overline{)85}$

7. $5 \overline{)31}$

8. $3 \overline{)23}$

Day 38

Division: Fill in the Blanks

Fill in the missing numbers in the division problems below.

1. $4\overline{)36}$ quotient: 9

2. $6\overline{)}$ quotient: 9

3. $4\overline{)}$ quotient: 15

4. $5\overline{)}$ quotient: 20

5. $3\overline{)}$ quotient: 2

6. $7\overline{)}$ quotient: 7

7. $2\overline{)}$ quotient: 5

8. $9\overline{)}$ quotient: 5

9. $8\overline{)}$ quotient: 8

10. $9\overline{)}$ quotient: 32

11. $7\overline{)}$ quotient: 9

12. $3\overline{)}$ quotient: 11

13. $6\overline{)}$ quotient: 9

14. $4\overline{)}$ quotient: 6

Day 39

Division Word Problems

1. Rachel bought three pairs of ballet shoes for $99. What is the cost of each pair of shoes?

____

2. Charlie has 21 kids in his class. If he divides the kids into 3 groups, how many kids will be in each group?

____

3. Sara loves her dolls; she has 12 of them. If she divides them into groups of 4, how many dolls will be in each group?

____

4. Harry has 72 toy trucks and cars. If he divides them into groups of 8, how many cars and trucks will be in each group?

____

DAY 40

2-DIGIT QUOTIENTS WITH REMAINDERS

Find the quotient of each problem below.

1.

```
     2 8 r 1
   _________
2 ) 5 7
  - 4
   ___
    1 7
  - 1 6
   ____
      1
```

2. $3\overline{)76}$

3. $4\overline{)65}$

4. $3\overline{)86}$

5. $2\overline{)37}$

6. $7\overline{)93}$

7. $5\overline{)73}$

8. $7\overline{)97}$

DAY 41

DIVIDING 3-DIGIT NUMBERS

Find the quotient of each problem below. Some may or may not have remainders

1.

```
     2 4 6 r 3
4 ) 9 8 7
  - 8
    1 8
  - 1 6
      2 7
    - 2 4
        3
```

2.

6) 2 9 7

3.

2) 5 9 2

4.

3) 7 6 9

5.

4) 8 6 7

6.

5) 9 3 4

Chapter 8: Geometry

Two-dimensional shapes can be classified by the number of sides they have. Polygons are considered **regular polygons** if all the sides are the same length. For example, a regular pentagon has 5 sides, and all of the sides are the same length. Side lengths that are the same are considered **congruent**. For example, a square has 4 congruent side lengths.

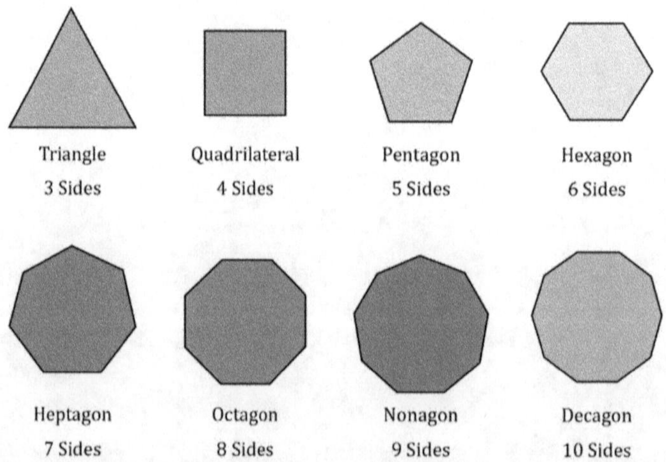

The most common three-dimensional figures include cones, cylinders, spheres, triangular and rectangular prisms, and cubes. When identifying and classifying three-dimensional figures it is important to be aware of the number of faces, edges, and vertices. **Faces** are the flat surfaces, **edges** are where two faces come together, and **vertices** are like corners, where more than two faces come together. For example, a rectangular prism has 6 faces, 12 edges, and 8 vertices.

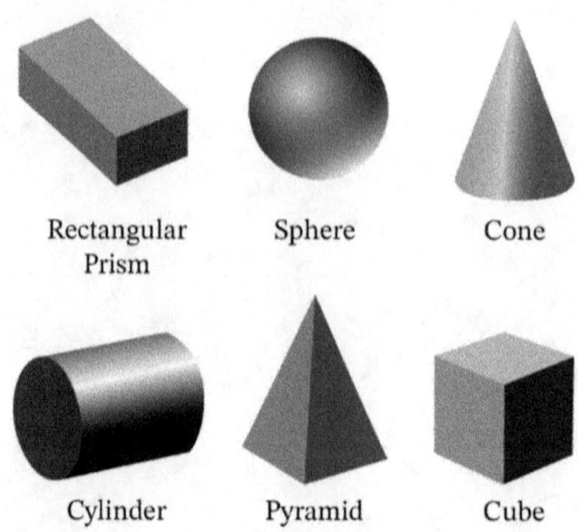

Perimeter is the distance around the outside of a two-dimensional shape. Perimeter can be calculated by finding the **sum** of a shape's **side lengths**.

The **area formula** for a **rectangle** is Area = Length × Width. This means the area can be determined by multiplying the number of **rows** by the number of **columns**. Think about the area of a rectangle as an array. For example, a rectangle that is 6 ft by 4 ft is essentially 6 columns and 4 rows. 6 groups of 4 is the same thing as 4 groups of 6, so the area can be calculated as 6×4 or 4×6. $6 \times 4 = 24$ and $4 \times 6 = 24$. The area is 24 square units.

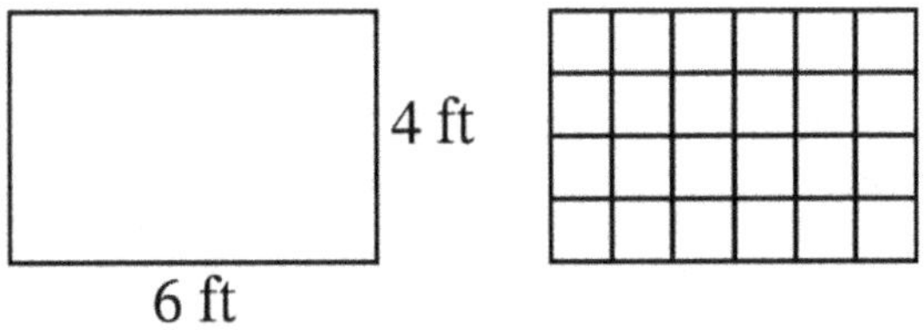

All three-dimensional figures occupy space. This space is defined as **volume**. Volume is also known as capacity. Spheres, prisms, pyramids, cylinders, cones, and cubes are common three-dimensional figures that have volume. The volume of a three-dimensional solid is measured in cubic units. For example, a cube has a length, width, and height. A cube that is 1 unit tall, 1 unit wide, and 1 unit deep is called a **unit cube**. The number of unit cubes that fills a solid determines its volume. The formula $V = l \times w \times h$, where l is length, w is width, and h is height, can be used to calculate the volume of a cube or rectangular prism.

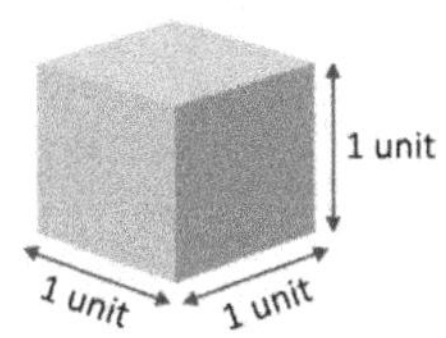

When four of these unit cubes are put together, the volume is 4 units^3.

A **point** is a fixed location in space, has no size or dimensions, and is commonly represented by a dot. A **line** is a set of points that extends infinitely in two opposite directions. It has length, but no width or depth. A line can be defined by any two distinct points that it contains. A **line segment** is a portion of a line that has definite endpoints. A **ray** is a portion of a line that extends from a single point on that line in one direction along the line. It has a definite beginning, but no ending.

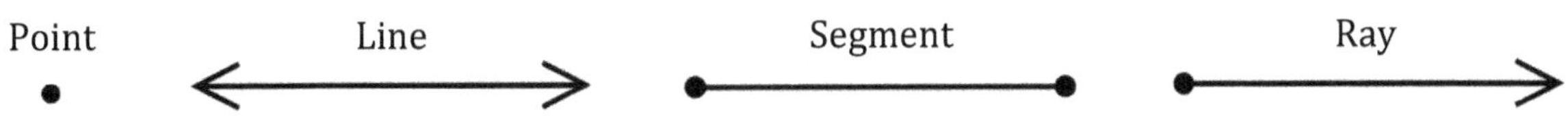

A **vertex** is formed when two straight lines or rays meet at a common endpoint.

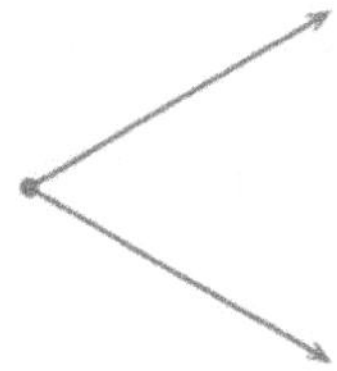

A **right angle** consists of 90 degrees and forms a corner. Angles that are less than 90 degrees are considered **acute angles.** Angles that are wider than 90 degrees but less than 180 degrees are considered **obtuse angles.**

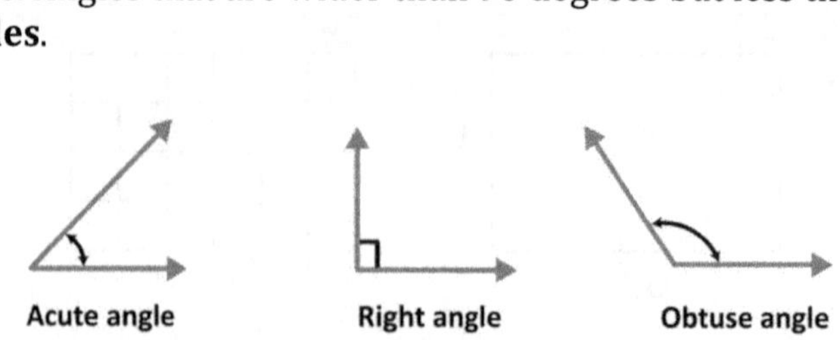

Triangles can be built from acute, obtuse, or right angles. For example, a triangle with only acute angles is referred to as an **acute triangle.** A triangle with an obtuse angle is called an **obtuse triangle.** A triangle containing a right angle is called a **right triangle.**

Triangle 1 contains a 90-degree angle, so it is classified as a right triangle. This can be verified using a protractor. Triangle 2 contains three acute angles, so it is classified as an acute triangle. Triangle 3 contains an obtuse angle, so it is an obtuse triangle.

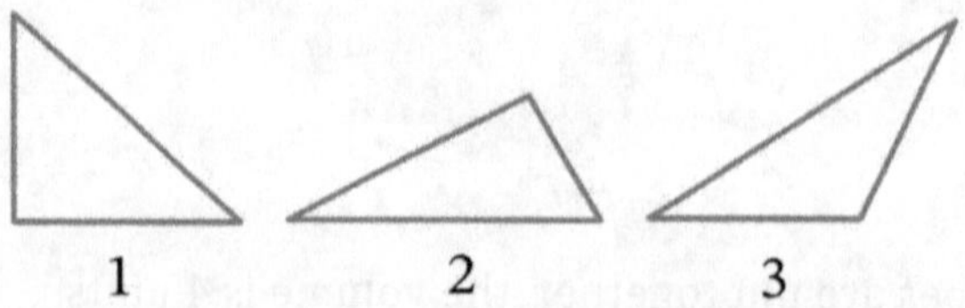

Intersecting lines are lines that have exactly one point in common. **Perpendicular lines** are lines that intersect at right angles. **Parallel lines** are lines in the same plane that have no points in common and never meet.

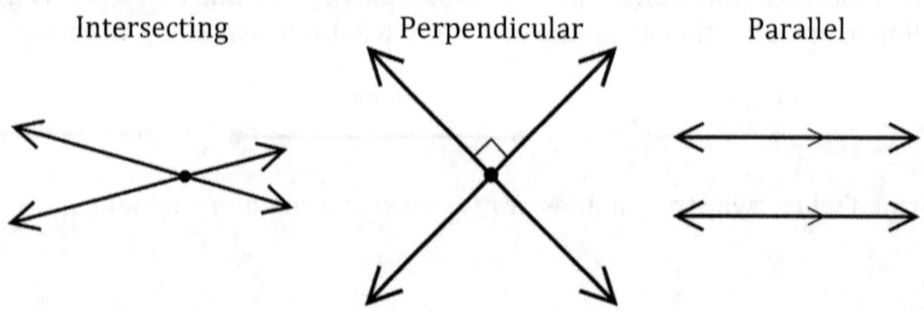

Complete one worksheet each day in this chapter. Each day should take no longer than 15 minutes to complete.

DAY 42

IDENTIFYING FLAT SHAPES

Write the number of sides each shape has, and then write the name of the shape.

1.

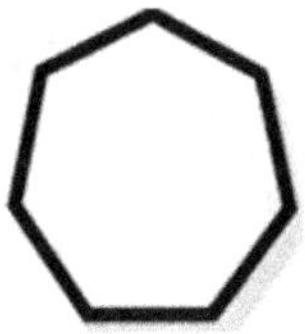

7 - Heptagon

2.

_____ - _____________

3.

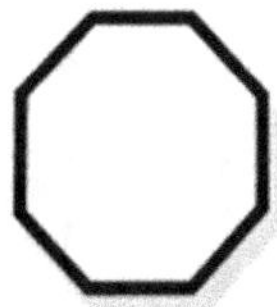

_____ - _____________

4.

_____ - _____________

5.

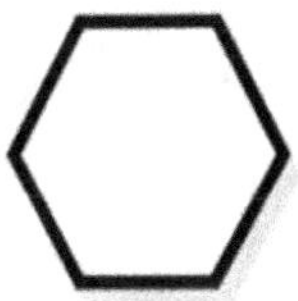

_____ - _____________

6.

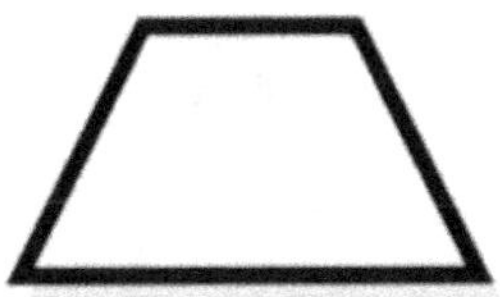

_____ - _____________

7.

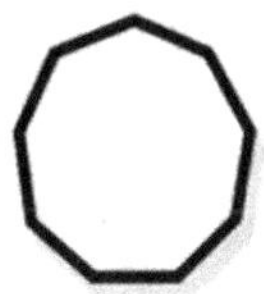

_____ - _____________

8.

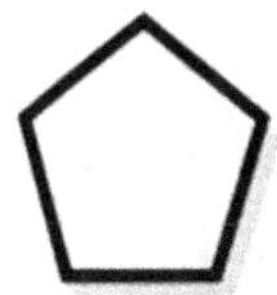

_____ - _____________

9.

_____ - _____________

10.

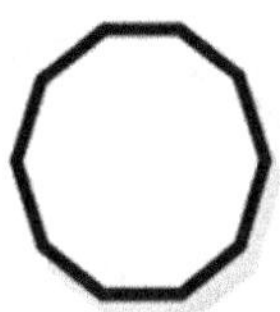

_____ - _____________

11.

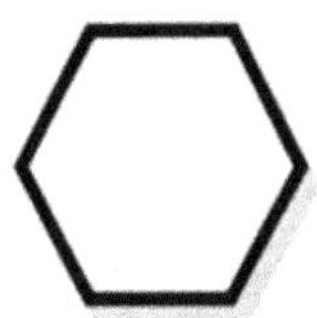

_____ - _____________

12.

_____ - _____________

13.

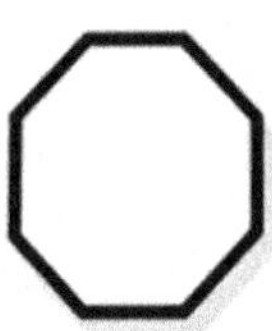

_____ - _____________

14.

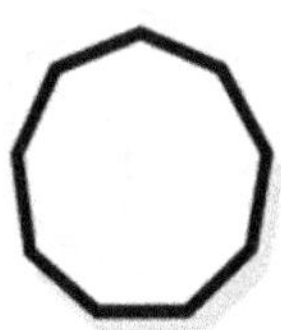

_____ - _____________

15.

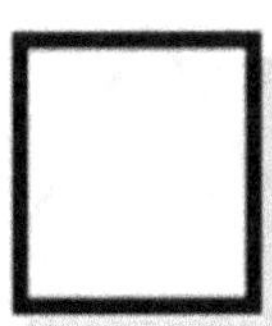

_____ - _____________

16.

_____ - _____________

Day 43

Identifying Solid Shapes

Identify the shapes and write the names in the blank space.

1.
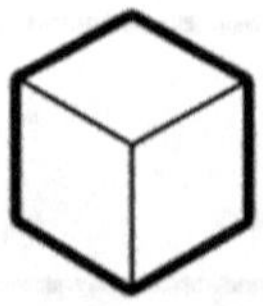

Cube

2.
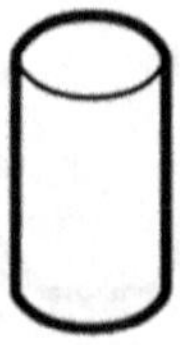

3.

4.
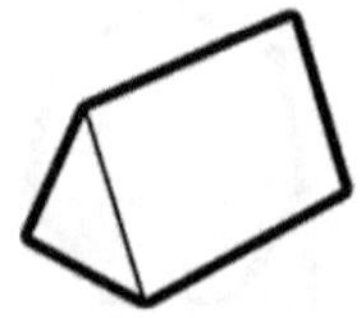

5.
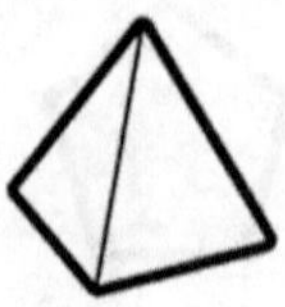

6.

7.
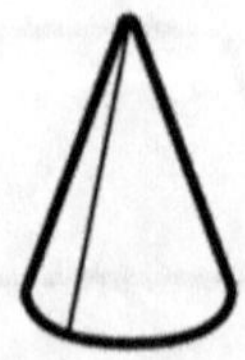

8.
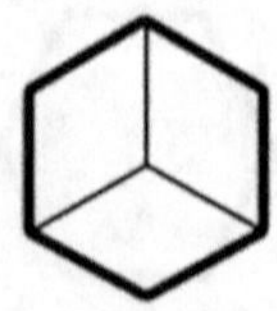

9.

10.
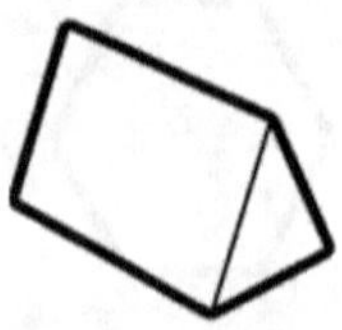

11.
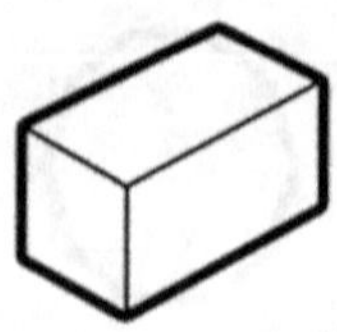

12.

13.
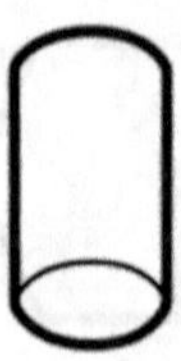

14.
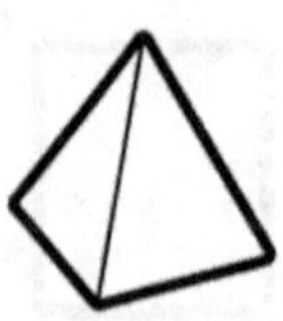

15.
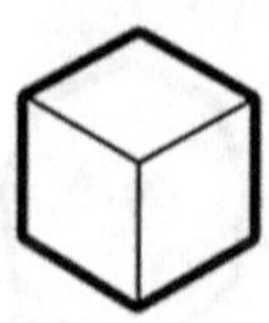

16.
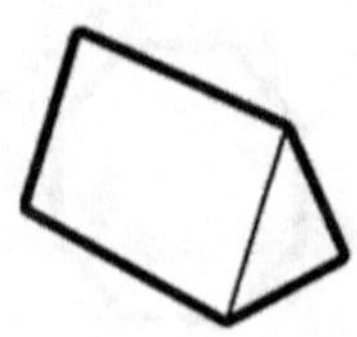

Day 44

Identifying Triangles

Name each triangle as equilateral, isosceles, or scalene.

1.

Isosceles

2.

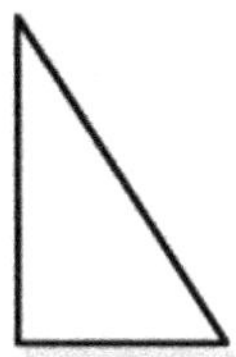

3.

4.

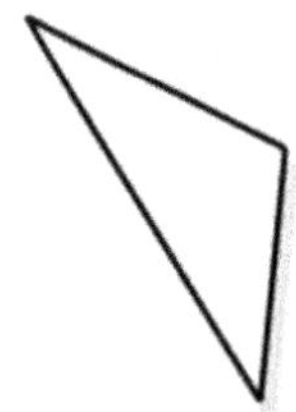

5.

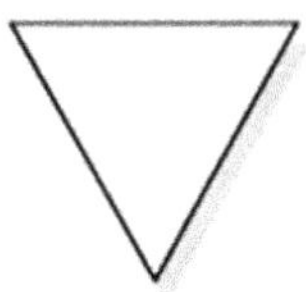

6.

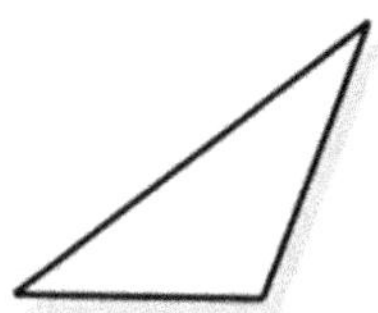

7.

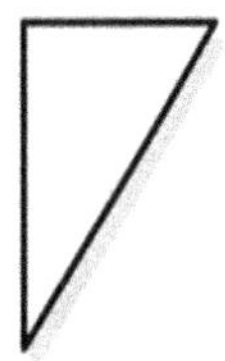

8.

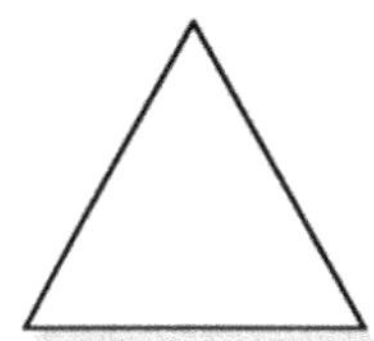

DAY 45

IDENTIFYING POINTS, LINES, AND RAYS

Identify each figure as a point, vertex, line, line segment, or ray.

1.

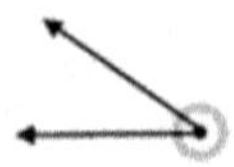

Vertex

2.

3.

4.

5.

6.

7.

8.

9.

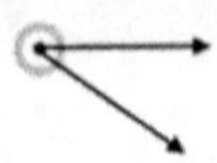

10.

11.

12.

Day 46

Identifying Parallel and Perpendicular Lines

Determine if each set of lines are parallel, perpendicular, or intersecting.

1.

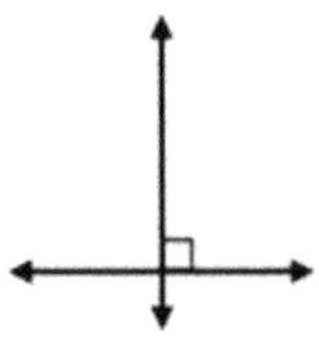

Perpendicular

2.

3.

4.

5.

6.

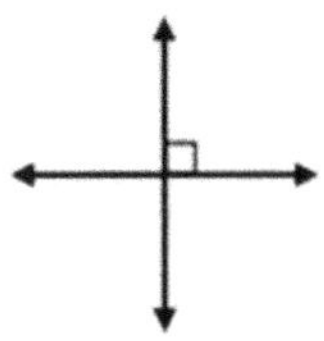

7.

8.

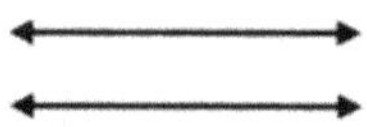

9.

10.

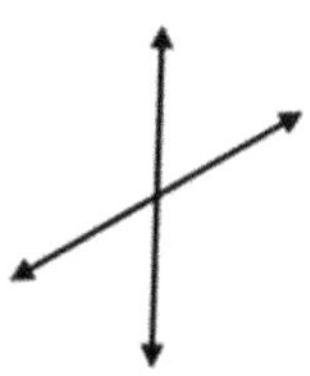

11.

12.

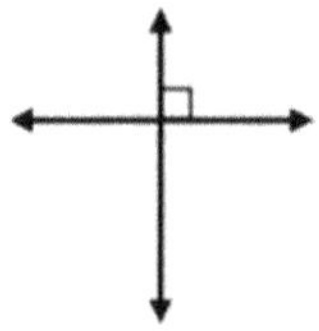

DAY 47

IDENTIFYING ANGLES

Determine if the angles are obtuse, acute, or right.

1.

Obtuse

2.

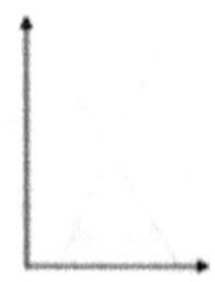

3.

4.

5.

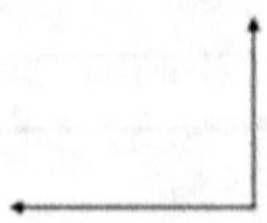

6.

7.

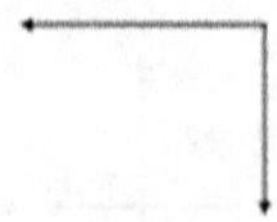

8.

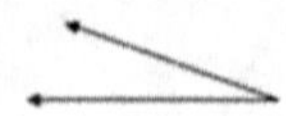

9.

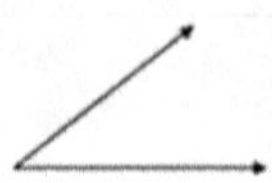

10.

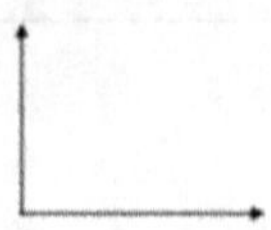

11.

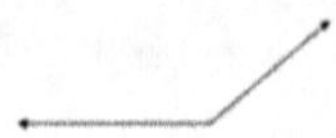

12.

DAY 48

FINDING AREA

Determine the area of each shape. Write the full problem out.

1.

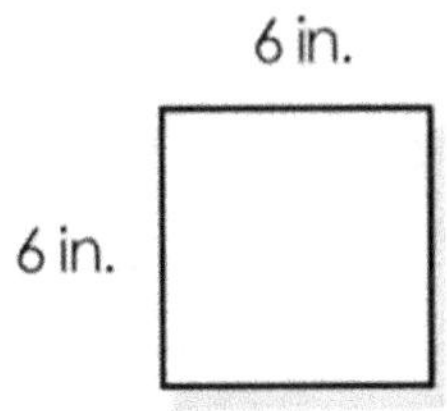

$6 \times 6 = 36\ \text{in}^2$

2.

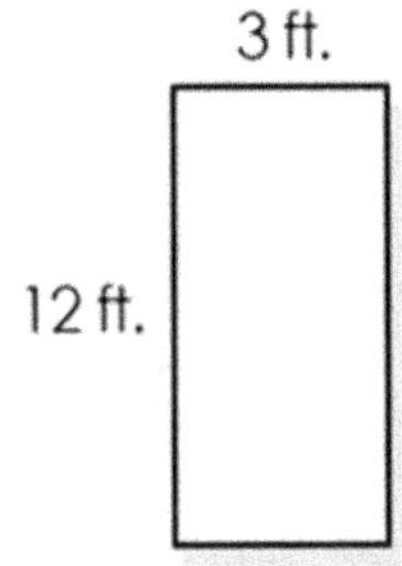

3.

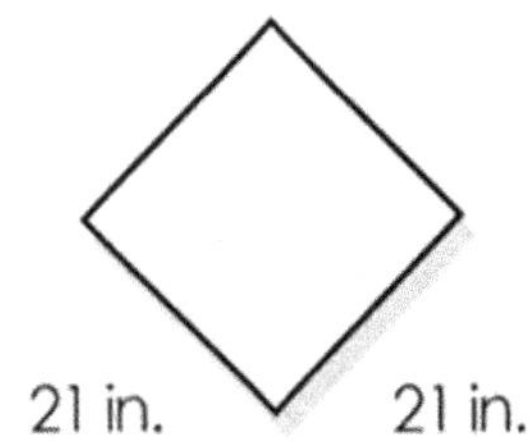

4.

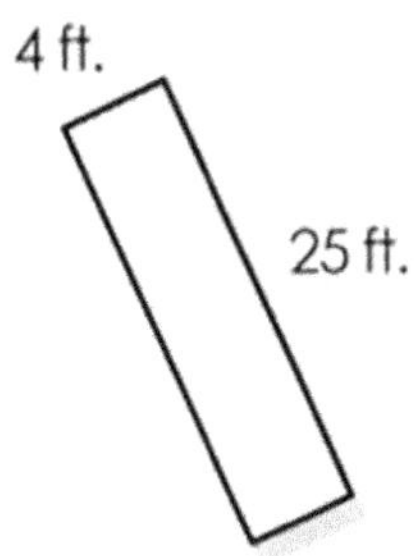

5.

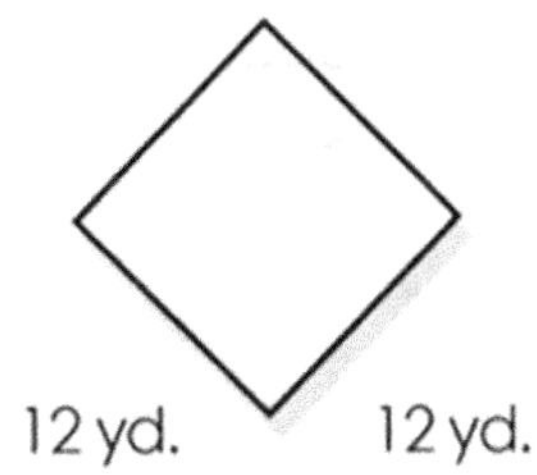

6.

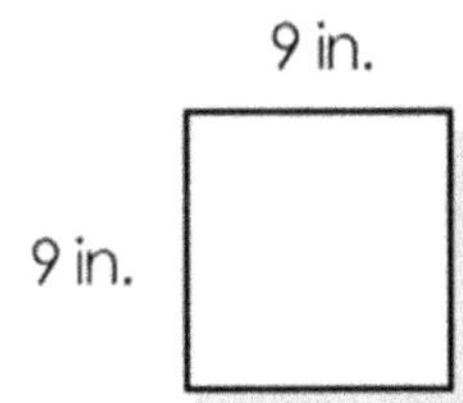

DAY 49

FINDING PERIMETER

Determine the perimeter of each shape. Write the full problem out.

1.

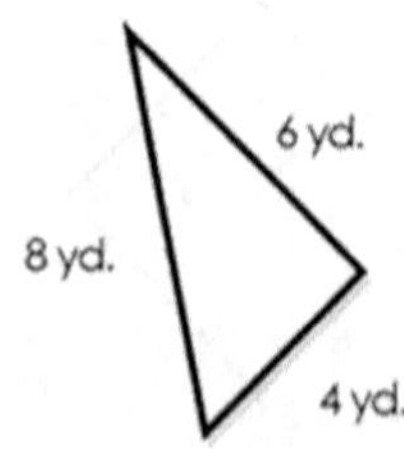

$8 + 4 + 6 = 18$ yd

2.

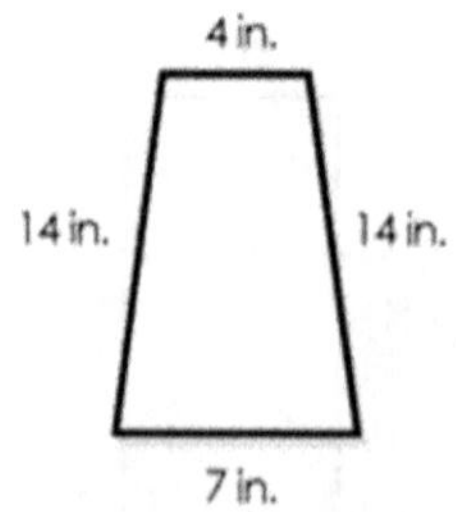

3.

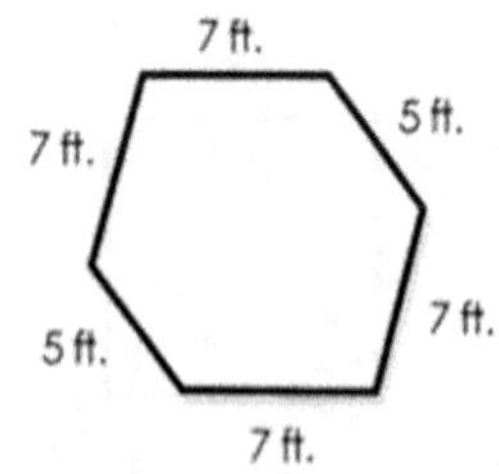

4.

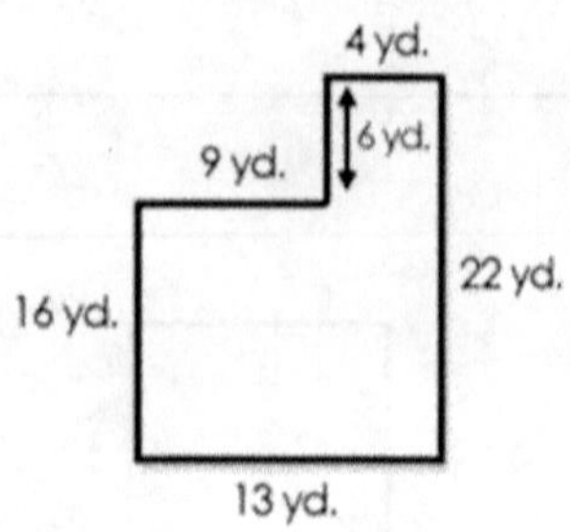

5.

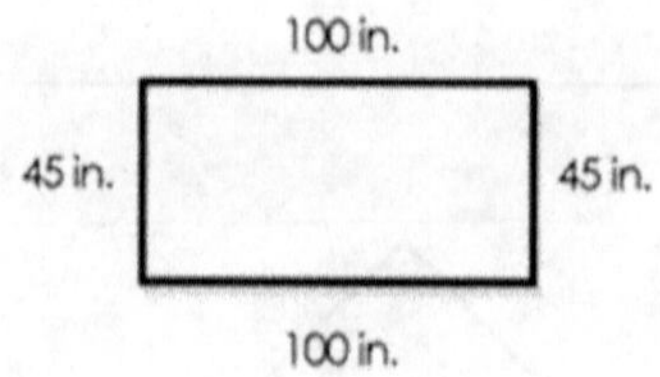

6.

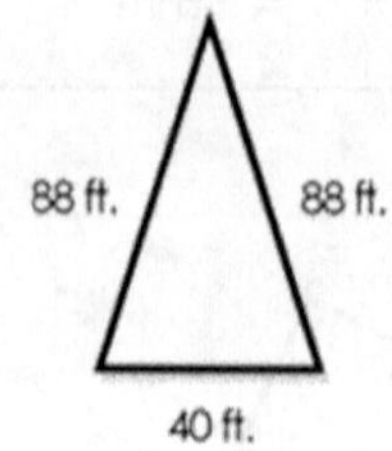

7.

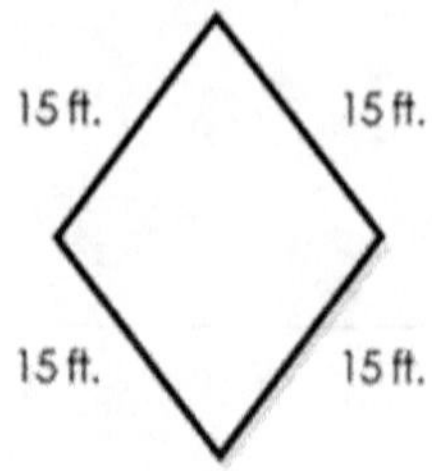

8.

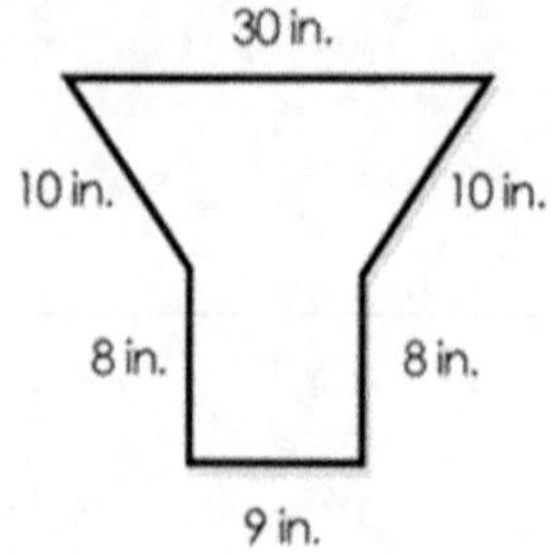

9.

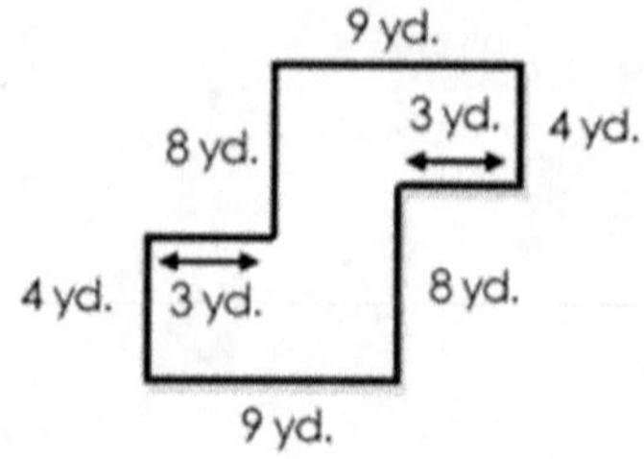

DAY 50

UNDERSTANDING VOLUME

Find the volume of the figures below.

1.

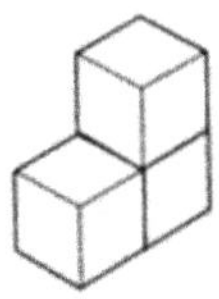

3 units^3

2.

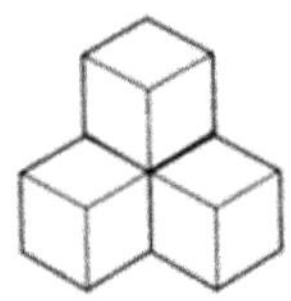

3.

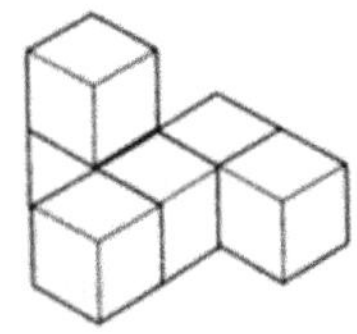

4.

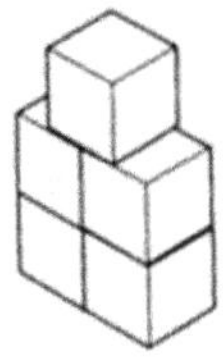

5.

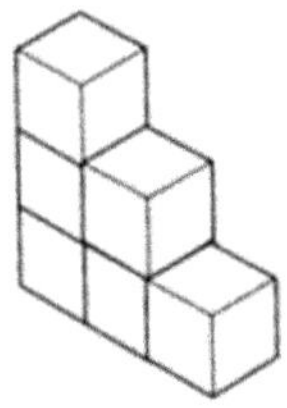

6.

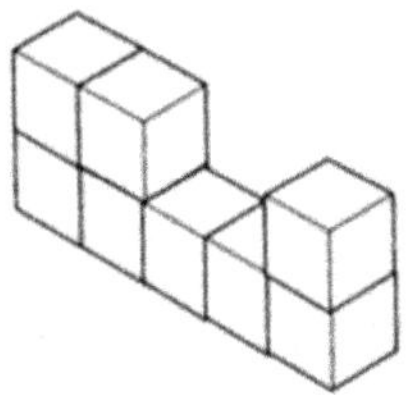

7.

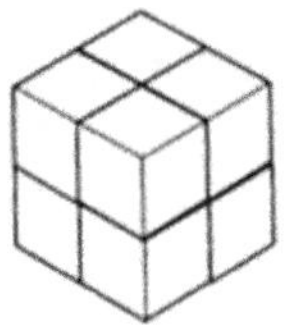

8.

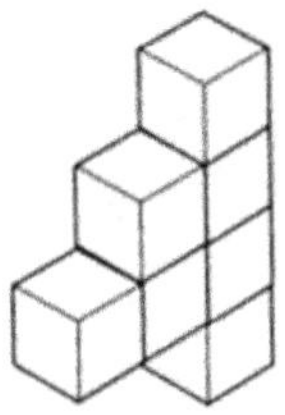

9.

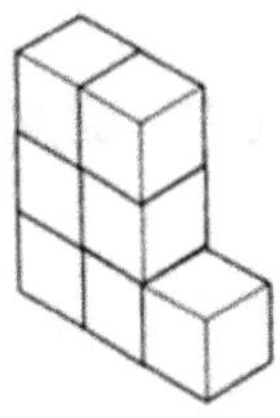

Workbook Answers

Chapter 1: Place Value & Number Sense

Day 1

Place Value: Big Numbers 1-1,000,000

1.	6 Millions	1 Hundred Thousands	3 Ten Thousands	8 Thousands	4 Hundreds	6 Tens	2 Ones
2.	3 Millions	1 Hundred Thousands	9 Ten Thousands	4 Thousands	6 Hundreds	7 Tens	5 Ones
3.	8 Millions	4 Hundred Thousands	1 Ten Thousands	7 Thousands	2 Hundreds	0 Tens	5 Ones
4.	2 Millions	7 Hundred Thousands	6 Ten Thousands	5 Thousands	4 Hundreds	4 Tens	7 Ones
5.	5 Millions	9 Hundred Thousands	2 Ten Thousands	5 Thousands	0 Hundreds	5 Tens	7 Ones

Day 2

Identifying Place Value: Word Problems

1. a. 3
   b. 4
   c. thousands
   d. tens
2. a. 3
   b. 2
   c. ten-thousands
   d. ones
3. a. 7
   b. 4
   c. tens
   d. ten-thousands
4. a. 9
   b. 1
   c. ones
   d. thousands
5. a. 9
   b. 5
   c. thousands
   d. ten-thousands

Day 3

Rounding Numbers

1. 3,000	10. 30,000
2. 4,000	11. 90,000
3. 9,000	12. 40,000
4. 6,000	13. 300,000
5. 1,000	14. 300,000
6. 8,000	15. 600,000
7. 40,000	16. 900,000
8. 80,000	17. 200,000
9. 20,000	18. 600,000

Day 4

Ordering up to 100,000

1.	100,280	100,289	100,381	100,891
2.	51,110	512,101	512,112	512,123
3.	855,622	856,628	856,629	865,622
4.	633,601	634,101	634,209	635,700
5.	20,115	21,115	231,115	231,150
6.	546,020	546,029	546,030	546,209
7.	355,111	355,891	375,375	375,787
8.	23,151	237,151	239,151	240,151
9.	145,280	145,381	146,891	155,289
10.	20,722	266,722	269,722	269,772

Chapter 2: Addition

Day 5

2-Digit 3-Row Addition: Regrouping

1. 73	2. 163	3. 135	4. 182
5. 98	6. 247	7. 141	8. 130
9. 157	10. 76	11. 178	12. 87

Day 6

Addition Word Problems

1. 64 apples

2. 56 minutes

3. 92 points

4. 173 minutes

Day 7

3-Digit Addition: Regrouping 1

1. 819	2. 1,309	3. 156	4. 756
5. 456	6. 253	7. 837	8. 1,221

Day 8

3-Digit Addition: Regrouping 2

1. 610	2. 444	3. 1,057	4. 718
5. 1,074	6. 1,062	7. 1,023	8. 1,600
9. 511	10. 853	11. 965	12. 767

Day 9

4-Digit Addition: Regrouping

1. 12,301	2. 10,761	3. 17,388	4. 5,201
5. 18,345	6. 3,656	7. 5,827	8. 8,464

Day 10

5-Digit Addition: Regrouping

1. 51,648	2. 100,997	3. 100,116	4. 40,099
5. 53,040	6. 144,064	7. 101,610	8. 62,008

Chapter 3: Subtraction

Day 11

2-Digit Subtraction: Borrowing

1. 29	2. 26	3. 79	4. 27
5. 47	6. 36	7. 23	8. 9

Day 12

Subtraction Word Problems

1. 32 shots
2. 44 cookies
3. 64 games
4. 24 snowflakes

Day 13

3-Digit Subtraction: Borrowing

1. 219	2. 23	3. 439	4. 489
5. 139	6. 266	7. 96	8. 158

Day 14

4-Digit Subtraction: Borrowing

1. 544	2. 1,091	3. 3,871	4. 2,779
5. 1,756	6. 6,487	7. 7,511	8. 2,264

Day 15

5-Digit Subtraction: Borrowing

1. 13,148	2. 15,814	3. 22,552	4. 4,247
5. 56,697	6. 28,313	7. 34,340	8. 36,202

Chapter 4: Decimals

Day 16

Adding Decimals

1. 3.59	2. 7.83	3. 16.11
4. 14.16	5. 10.43	6. 74.36

Day 17

Adding Money

1. $5.03	2. $11.63	3. $9.22
4. $6.40	5. $12.75	6. $22.27
7. $22.53	8. $19.64	9. $15.13

Day 18

Adding Decimals Word Problems

1. 52.64 ounces
2. 7.2 miles
3. 38.28 minutes
4. 15 pounds

Day 19

Rounding Decimals

1.	93	11.	34
2.	13	12.	32
3.	33	13.	77
4.	25	14.	72
5.	58	15.	12
6.	65	16.	44
7.	83	17.	46
8.	42	18.	56
9.	69	19.	21
10.	20	20.	19

Day 20

Subtracting Decimals

1. 2.16	2. 34.62	3. 4.65
4. 326.28	5. 854.43	6. 68.99

Day 21

Subtracting Money

1. \$2.47	2. \$3.14	3. \$3.18
4. \$1.86	5. \$1.15	6. \$1.85
7. \$4.79	8. \$22.28	9. \$13.54

Day 22

Subtracting Decimals Word Problems

1. \$12.18

2. \$3.21

3. \$3.18

4. \$30.10

Chapter 5: Fractions

Day 23

Writing Fractions

1. $\frac{1}{3}$
2. $\frac{2}{3}$
3. $\frac{1}{4}$
4. $\frac{2}{4}$ or $\frac{1}{2}$
5. $\frac{3}{8}$
6. $\frac{4}{8}$ or $\frac{1}{2}$
7. $\frac{3}{16}$
8. $\frac{9}{16}$

Day 24

Comparing Fractions

1. $>$
2. $<$
3. $>$
4. $<$
5. $>$
6. $<$
7. $<$
8. $>$
9. $>$
10. $>$
11. $=$
12. $>$
13. $>$
14. $<$
15. $<$
16. $<$
17. $<$
18. $>$
19. $<$
20. $>$

Day 25

Ordering Mixed Numbers

1. $1\frac{1}{10}, 1\frac{6}{10}, 2\frac{1}{10}, 2\frac{3}{10}$

2. $4\frac{4}{8}, 6\frac{2}{8}, 9\frac{3}{8}, 9\frac{6}{8}$

3. $6\frac{1}{4}, 6\frac{2}{4}, 7\frac{1}{4}, 7\frac{3}{4}$

4. $8\frac{1}{9}, 8\frac{2}{9}, 8\frac{3}{9}, 9\frac{1}{9}$

5. $1\frac{1}{12}, 2\frac{3}{12}, 3\frac{1}{12}, 3\frac{8}{12}$

6. $5\frac{1}{54}, 5\frac{2}{54}, 5\frac{3}{54}, 5\frac{7}{54}$

Day 26

Adding Fractions with Common Denominators

1. $\frac{4}{5}$
2. $\frac{8}{10}$
3. $\frac{5}{6}$
4. $\frac{10}{12}$
5. $\frac{8}{9}$
6. $\frac{8}{14}$
7. $\frac{5}{7}$
8. $\frac{3}{4}$
9. $\frac{5}{6}$
10. $\frac{7}{8}$
11. $\frac{9}{11}$
12. $\frac{12}{15}$
13. $\frac{12}{14}$
14. $\frac{2}{3}$
15. $\frac{6}{8}$
16. $\frac{4}{5}$
17. $\frac{11}{13}$
18. $\frac{7}{10}$

Day 27

Subtracting Fractions with Common Denominators

1. $\frac{4}{15}$
2. $\frac{5}{9}$
3. $\frac{5}{14}$
4. $\frac{2}{8}$
5. $\frac{9}{39}$
6. $\frac{13}{39}$
7. $\frac{7}{10}$
8. $\frac{4}{15}$
9. $\frac{10}{16}$
10. $\frac{2}{7}$
11. $\frac{2}{26}$
12. $\frac{19}{62}$
13. $\frac{9}{24}$
14. $\frac{29}{54}$
15. $\frac{83}{98}$

Day 28

Simplifying Fractions

1. $\frac{3}{3}, \frac{1}{3}$
2. $\frac{4}{4}, \frac{1}{3}$
3. $\frac{3}{3}, \frac{2}{5}$
4. $\frac{2}{2}, \frac{4}{5}$
5. $\frac{5}{5}, \frac{2}{5}$
6. $\frac{5}{5}, \frac{4}{5}$

Chapter 6: Multiplication

Day 29

Multiplication Table Practice

1.

×	**1**	**2**	**3**	**4**	**5**
6	6	12	18	24	30
7	7	14	21	28	35
8	8	16	24	32	40

2.

×	**7**	**8**	**9**
4	28	32	36
5	35	40	45
6	42	48	54
7	49	56	63
8	56	64	72

3.

×	**9**	**10**	**11**
7	63	70	77
8	72	80	88
9	81	90	99
10	90	100	110
11	99	110	121

4.

×	**2**	**3**	**4**	**5**	**6**
20	40	60	80	100	120
25	50	75	100	125	150
27	54	81	108	135	162

Day 30

Mixed Multiplication Facts

1. 0	2. 16	3. 1	4. 9	5. 10	6. 12	7. 27
8. 20	9. 6	10. 35	11. 4	12. 24	13. 42	14. 24
15. 18	16. 25	17. 7	18. 14	19. 8	20. 10	21. 16
22. 18	23. 36	24. 9	25. 45	26. 9	27. 48	28. 81

Day 31

Multiplying 1-Digit Numbers by 2-Digit Numbers

1. 48
2. 93
3. 86
4. 77
5. 55
6. 88
7. 15
8. 108
9. 129
10. 68

Day 32

Multiplying by 10s

1. 260
2. 710
3. 670
4. 220
5. 380
6. 840
7. 190
8. 180
9. 940
10. 290

Day 33

Multiplying 1-Digit Numbers by 2-Digit Numbers: Regrouping

1. 92
2. 406
3. 172
4. 261
5. 180
6. 220
7. 288
8. 495
9. 198
10. 354

Day 34

Multiplying 1-Digit Number by 3-Digit Numbers: Regrouping

1. 568
2. 1,326
3. 1,023
4. 847
5. 4,188
6. 1,008
7. 414
8. 478
9. 1,624
10. 1,128
11. 1,060
12. 486

Day 35

Multiplying 2-Digit Numbers by 2-Digit Numbers: Regrouping

1. 735
2. 156
3. 1,188
4. 3,240
5. 950
6. 4,620
7. 2,553
8. 667
9. 4,424
10. 5,544

Chapter 7: Division

Day 36

Basic Division

1. 3
2. 8
3. 6
4. 4
5. 9
6. 7
7. 3
8. 8
9. 3
10. 3
11. 9
12. 10

Day 37

Division with Remainders

1. 6 r 1
2. 5 r 1
3. 4 r 1
4. 4 r 7
5. 4 r 2
6. 9 r 4
7. 6 r 1
8. 7 r 2

Day 38

Division Fill in the Blanks

1. 36
2. 54
3. 60
4. 100
5. 6
6. 49
7. 10
8. 45
9. 64
10. 288
11. 63
12. 33
13. 54
14. 24

Day 39

Division Word Problems

1. $33

2. 7 kids

3. 3 dolls

4. 9 cars and trucks

Day 40

2-Digit Quotients with Remainders

1. 28 r 1
2. 24 r 4
3. 16 r 1
4. 28 r 1
5. 18 r 1
6. 13 r 2
7. 14 r 3
8. 13 r 6

Day 41

Dividing 3-Digit Numbers

1. 246 r 3
2. 49 r 3
3. 296
4. 256 r 1
5. 216 r 3
6. 186 r 4

Chapter 8: Geometry

Day 42

Identifying Flat Shapes

1.	7 – heptagon	2.	4 – square	3.	8 – octagon	4.	4 – rectangle
5.	6 – hexagon	6.	4 – trapezoid	7.	9 – nonagon	8.	5 – pentagon
9.	4 – rectangle	10.	10 – decagon	11	6 – hexagon	12.	3 – triangle
13.	8 – octagon	14.	9 – nonagon	15.	4 – square	16.	6 – hexagon

Day 43

Identifying Solid Shapes

1. cube
2. cylinder
3. sphere
4. triangular prism
5. pyramid
6. rectangular prism
7. cone
8. cube
9. sphere
10. triangular prism
11. rectangular prism
12. cone
13. cylinder
14. pyramid
15. cube
16. triangular prism

Day 44

Identifying Triangles

1. isosceles
2. scalene
3. isosceles
4. scalene
5. equilateral
6. scalene
7. scalene
8. equilateral

Day 45

Identifying Points, Lines, and Rays

1. vertex
2. ray
3. line segment
4. point
5. ray
6. line segment
7. vertex
8. line
9. vertex
10. point
11. line
12. line segment

Day 46

Identifying Parallel and Perpendicular Lines

1. perpendicular
2. parallel
3. intersecting
4. parallel
5. intersecting
6. perpendicular
7. intersecting
8. parallel
9. perpendicular
10. intersecting
11. parallel
12. perpendicular

Day 47

Identifying Angles

1. obtuse
2. right
3. acute
4. obtuse
5. right
6. obtuse
7. right
8. acute
9. acute
10. right
11. obtuse
12. acute

Day 48

Finding Area

1. 36 sq. in
2. 36 sq. ft
3. 441 sq. in
4. 100 sq. ft
5. 144 sq. yd
6. 81 sq. in

Day 49

Finding Perimeter

1. 18 yd
2. 39 in
3. 38 ft
4. 70 yd
5. 290 in
6. 216 ft
7. 60 ft
8. 75 in
9. 48 yd

Day 50

Understanding Volume

1. 3 units^3
2. 4 units^3
3. 6 units^3
4. 5 units^3
5. 6 units^3
6. 8 units^3
7. 8 units^3
8. 7 units^3
9. 7 units^3

www.ingramcontent.com/pod-product-compliance
Lightning Source LLC
LaVergne TN
LVHW082358200926
848481LV00012B/3610